SpringerBriefs in Molecular Science

Ultrasound and Sonochemistry

Series editors

Bruno G. Pollet, Faculty of Engineering, Norwegian University of Science
and Technology, Trondheim, Norway
Muthupandian Ashokkumar, School of Chemistry, The University of Melbourne,
Parkville, VIC, Australia

SpringerBriefs in Molecular Science: Ultrasound and Sonochemistry is a series of concise briefs that present those interested in this broad and multidisciplinary field with the most recent advances in a broad array of topics. Each volume compiles information that has thus far been scattered in many different sources into a single, concise title, making each edition a useful reference for industry professionals, researchers, and graduate students, especially those starting in a new topic of research.

More information about this series at http://www.springer.com/series/15634

About the Series Editors

Bruno G. Pollet is Full Professor of Renewable Energy at the Norwegian University of Science and Technology (NTNU). He is also Visiting Professor at the University of Ulster (UK). He was a Visiting Professor at the University of Yamanashi (Japan) as well as Chief Technology Officer at Power and Water (KP2M Ltd, UK) designing and developing energy storage and water purification systems. He was previously Head of R&D at Coldharbour Marine Ltd (UK) working in the area of water treatment/disinfection. He was awarded Diploma in Chemistry and Material Sciences from the Université Joseph Fourier (Grenoble, France), B.Sc. (Hons) in Applied Chemistry from the Coventry University (UK) and M.Sc. in Analytical Chemistry from The University of Aberdeen (UK). He also gained his Ph.D. in Physical Chemistry in the field of Electrochemistry and Sonochemistry (Sonoelectrochemistry) under the supervision of Profs. J. Phil Lorimer and T. J. Mason at the Sonochemistry Centre, Coventry University (UK). He has published many scientific publications, articles and books (including three books) in the field of Sonochemistry, Fuel Cells, Electrocatalysis and Electrochemical Engineering (over 200 publications so far).

Professor Muthupandian Ashokkumar is Physical Chemist who specializes in Sonochemistry, teaches undergraduate and postgraduate Chemistry and is a senior academic staff member of the School of Chemistry, The University of Melbourne. He is a renowned sonochemist who has developed a number of novel techniques to characterize acoustic cavitation bubbles and has made major contributions of applied sonochemistry to the materials, food and dairy industry. His research team has developed a novel ultrasonic processing technology for improving the functional properties of dairy ingredients. Recent research also involves the ultrasonic synthesis of functional nano- and bioma-terials that can be used in energy production, environmental remediation and diagnostic and therapeutic medicine. He is the Editor-in-Chief of Ultrasonics Sonochemistry, an international journal devoted to sonochemistry research with a Journal Impact Factor of 4.8. He has edited/co-edited several books and special issues for journals, published ~ 350 refereed papers (H-Index: 45) in high-impact international journals and books and delivered over 150 invited/keynote/plenary lectures at international conferences and academic institutions. He is the recipient of several prizes, awards and fellowships, including the Grimwade Prize in Industrial Chemistry. He is a Fellow of the RACI since 2007.

Thomas Seak Hou Leong
Sivakumar Manickam · Gregory J. O. Martin
Wu Li · Muthupandian Ashokkumar

Ultrasonic Production of Nano-emulsions for Bioactive Delivery in Drug and Food Applications

 Springer

Thomas Seak Hou Leong
ARC Dairy Innovation Hub
The University of Melbourne
Parkville, VIC
Australia

and

School of Chemistry
The University of Melbourne
Parkville, VIC
Australia

Sivakumar Manickam
Nanotechnology and Advanced
    Materials (NATAM), Faculty
    of Engineering
University of Nottingham Malaysia
    Campus
Semenyih
Malaysia

Gregory J. O. Martin
ARC Dairy Innovation Hub
The University of Melbourne
Parkville, VIC
Australia

and

Department of Chemical Engineering
    and Biomolecular Engineering
The University of Melbourne
Parkville, VIC
Australia

Wu Li
ARC Dairy Innovation Hub
The University of Melbourne
Parkville, VIC
Australia

and

School of Chemistry
The University of Melbourne
Parkville, VIC
Australia

Muthupandian Ashokkumar
ARC Dairy Innovation Hub
The University of Melbourne
Parkville, VIC
Australia

and

School of Chemistry
The University of Melbourne
Parkville, VIC
Australia

ISSN 2191-5407   ISSN 2191-5415   (electronic)
SpringerBriefs in Molecular Science
ISSN 2511-123X   ISSN 2511-1248   (electronic)
SpringerBriefs in Ultrasound and Sonochemistry
ISBN 978-3-319-73490-3   ISBN 978-3-319-73491-0   (eBook)
https://doi.org/10.1007/978-3-319-73491-0

Library of Congress Control Number: 2018930370

Printed on acid-free paper

This Springer imprint is published by the registered company Springer International Publishing AG part of Springer Nature
The registered company address is: Gewerbestrasse 11, 6330 Cham, Switzerland

Preface

Emulsions are a semi-stable mixture of two immiscible liquids, one of which is dispersed as droplets within a continuous phase of the other. Emulsions serve an important role in enhancing the bioavailability of fats, lipids and nutrients in both aqueous and organic systems and hence are found across a wide range of products in food and pharmaceutical applications.

There is great interest in the use of ultrasound to produce emulsions. Ultrasound is a technology that can create emulsions relatively efficiently and effectively compared to other techniques such as rotor–stator mixing, high-pressure homogenization and microfluidization. The interaction of ultrasound with hydrocolloids and biopolymers that are often used to stabilize emulsions can offer advantages such as improved stability or greater control of the size distribution of droplets formed. This SpringerBrief will provide an overview of ultrasonic emulsification (Chap. 1), guide towards the most suitable parameters required for effective ultrasonic emulsion formation (Chap. 2) and showcase recent applications in which stable emulsions produced from ultrasound have been used to develop novel drug formulations and functional foods (Chap. 3).

Parkville, Australia
Semenyih, Malaysia
Parkville, Australia
Parkville, Australia
Parkville, Australia

Thomas Seak Hou Leong
Sivakumar Manickam
Gregory J. O. Martin
Wu Li
Muthupandian Ashokkumar

Acknowledgements

Authors Leong, Martin, Li and Ashokkumar acknowledge research support under the Australian Research Council's Industrial Transformation Research Program (ITRP) funding scheme (project number IH120100005). The ARC Dairy Innovation Hub is a collaboration between The University of Melbourne, The University of Queensland and Dairy Innovation Australia Ltd. Thomas Leong also wishes to acknowledge the Dyason Fellowship provided by The University of Melbourne for travel support to visit the University of Nottingham.

Acknowledgements

Contents

Glossary of Key Terms

Emulsifier	An amphiphilic molecule that assembles at the interface of oil/water phase boundaries, reducing the surface tension, thereby resisting phase separation of the dispersed droplets. Additional mechanisms conferring stability are the provision of steric and/or electrostatic barriers to inter-droplet interaction.
Stabilizer	Surface active macromolecules that are added to increase the viscosity of the continuous phase of an emulsion in order to reduce the mobility of dispersed droplets, slowing down inter-droplet collisions that may lead to droplet coalescence.
Acoustic cavitation	The nucleation, growth and collapse of gas nuclei in fluids due to the application of an oscillating sound field. The collapse event is accompanied by the release of a large amount of energy in the form of pressure shock waves, fluid streaming and microjets as well as temperature hot spots.
Emulsion droplet size (EDS)	The characteristic size of the dispersed phase droplets of an emulsion.
Coalescence	The combination of two droplets to form a larger droplet.
Ostwald Ripening	The process by which molecules from small emulsion droplets diffuse through the continuous phase to larger droplets.
Continuous phase	The bulk phase of the emulsion.
Dispersed phase	The phase which is dispersed in the emulsion in the form of droplets. The droplets of the dispersed phase are stabilized with a coating of emulsifier.

Immiscible liquids   Two liquids that are not soluble, or barely soluble, within each other. For example, oil is barely soluble within water, and vice versa.

Micelle   A self-assembled aggregate of surface active molecules that are dispersed as a colloidal suspension. A typical structure in aqueous medium is a sphere with surfactant molecules aligned in such a way so that hydrophilic head groups point towards the solvent, while hydrophobic tail groups are pointed towards the centre.

Chapter 1
Introduction

Abstract An emulsion is a dispersion of two immiscible liquids, combined together to form a semi-stable mixture. This chapter will briefly introduce the main classes of emulsions and the mechanisms of ultrasound that enable efficient production of emulsions.

Keywords Emulsion · Ultrasound · Emulsion droplet size · Cavitation
Microjets · Shock wave

The immiscible liquid phases in an emulsion generally consist of an organic (oil) phase and an aqueous (water) phase. Two main classes of emulsions are the water-in-oil (W/O) and the oil-in-water (O/W) emulsion. Other classes such as multiple emulsions of W/O/W (water droplets dispersed in oil droplets dispersed in a continuous water phase) or O/W/O type (oil droplets dispersed in water droplets dispersed in a continuous oil phase) are also possible and are of interest due to their ability to encapsulate and protect bioactive materials. The common classes of emulsions are depicted in Fig. 1.1.

Emulsions are inherently thermodynamically unstable dispersions that will phase separate (become two separate, continuous phases, e.g. oil over water) over time due to thermal and kinetic instability. An emulsion product intended for consumers should not phase separate within its usable shelf life. To stabilize an emulsion requires use of a suitable emulsifier that will interface at the boundary between the two phases, thereby reducing the interfacial tension. An emulsifier is usually a surface active molecule (surfactant) with amphiphilic properties consisting of a hydrophilic (water loving) end and a hydrophobic (water fearing) hydrocarbon end. The hydrophilic end will become positioned in the aqueous phase, while the hydrocarbon chain will become positioned in the nonpolar/organic phase (Fig. 1.2). The relative size of the hydrophilic and hydrophobic (lipophilic) sections of the surfactant molecule is commonly referred to as the hydrophilic/lipophilic balance (HLB). The HLB of the surfactant will influence whether an emulsion is O/W or W/O. Low HLB emulsifiers (HLB = 1 − 4) favour the formation of W/O emulsions, while higher HLB emulsifiers (7–12) favour the formation of O/W emulsions [1].

© The Author(s), under exclusive licence to Springer International
Publishing AG, part of Springer Nature 2018
T. S. H. Leong et al., *Ultrasonic Production of Nano-emulsions for Bioactive Delivery in Drug and Food Applications*, Ultrasound and Sonochemistry, https://doi.org/10.1007/978-3-319-73491-0_1

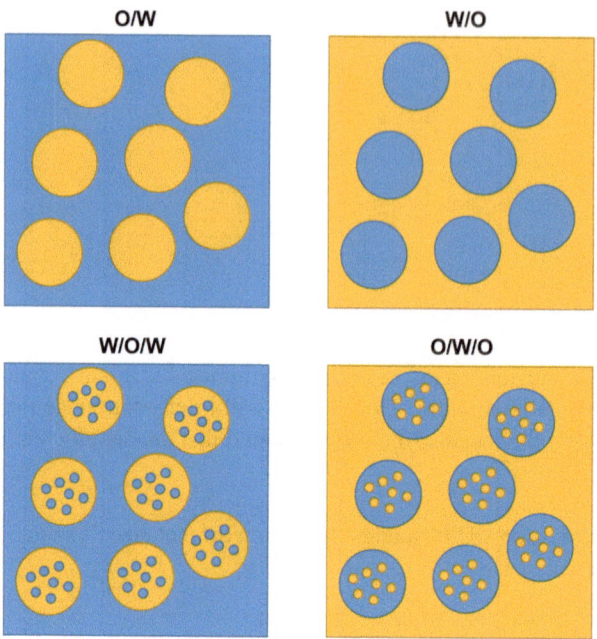

Fig. 1.1 Schematic representations of common types of simple and multiple (double) emulsions. Oil (O) is yellow, and water (W) is blue

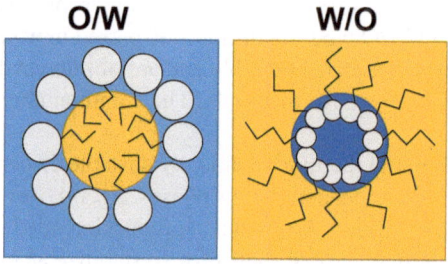

Fig. 1.2 Schematic depiction of the stabilization of an oil-in-water (O/W) or water-in-oil (W/O) emulsion droplet with surface active molecules (surfactants). The O/W emulsion is stabilized by a surfactant with a high hydrophilic/lipophilic balance (HLB) (large hydrophilic section/small lipophilic section). The W/O emulsion is stabilized by a surfactant with a low HLB (small hydrophilic section/large lipophilic section)

Emulsifiers can be synthetically created, e.g. from fats and oils, or they may be derived from naturally occurring surface active biocomponents such as proteins (e.g. milk proteins, vegetable proteins) and hydrocolloids (e.g. polysaccharides derived from plant material). The role of the emulsifier is to prevent spontaneous coalescence of the dispersed phase, by reducing the interfacial tension (which reduces the thermodynamic driving force for phase separation) and by creating physical or electrostatic barriers preventing droplets from coalescing (which provides kinetic stability).

In the food and pharmaceutical industries, emulsions are an important vehicle for the delivery of bioactive materials. In particular, stable O/W type emulsions with the oil phase dispersed as small droplets can be used to efficiently deliver oil-soluble components or nutritionally beneficial oils into foods [2–4]. These can be natural (e.g. cows' milk) or artificial (e.g. mayonnaise and sauces). The bulk phase of O/W emulsions being of an aqueous origin makes them more readily palatable to consumers. By the same principle, W/O/W double emulsions can be used to stably encapsulate an aqueous phase component within an oil droplet phase [5]. For food applications, O/W and W/O/W emulsions can be delivered in the form of functional drinks and beverages. W/O and O/W/O emulsions are well suited for cosmetics and application on skin, as the oily continuous phase can provide good moisturizing properties. These emulsions are also commonly found in foods in the form of spreads such as butters and margarines.

1.1 Emulsion Droplet Size

One of the important attributes that governs emulsion stability, appearance and taste is the emulsion droplet size (EDS) and size distribution. Based on the EDS, emulsions can be categorized into three general size groups—macroemulsions (0.5–100 μm), mini (nano)emulsions (100–1000 nm) and microemulsions (10–100 nm) [6]. For simplicity, in this chapter we will refer to all sub-micron emulsions as nanoemulsions.

In food products, the visual appearance is often an important consideration for consumers. Macrosized emulsions are characterized by a 'milky' opaque appearance. While this may be desirable for some products (e.g. milk and sauces), an opaque appearance may be deemed as undesirable for other products such as liquid beverages. This can be overcome if nanoemulsions are made in which the EDS is smaller than ~100 nm [7], as the reduced size of the oil droplets alters the light diffraction behaviour of the droplets, rendering the emulsions to appear translucent and clear (Fig. 1.3a).

Emulsions with an EDS smaller than ~100 nm become kinetically stable to creaming [6], as the Brownian motion of the droplets overcomes the natural buoyancy of individual droplets. Colloidal processes such as Ostwald Ripening [9] and coalescence will result in eventual phase separation. While these processes are characteristically slow for emulsions with these small droplet sizes, they occur more rapidly at higher droplet concentrations. However, the use of surfactants that provide steric or electrostatic repulsion can slow coalescence to the extent that minimal phase separation will occur within the usable shelf life.

The rheological properties of an emulsion are also strongly influenced by the EDS. As an emulsion decreases in droplet size, the number of droplets present in the fluid increases for a given concentration of the dispersed phase. This results in more particle–particle interactions and hence an increased resistance to flow; i.e.,

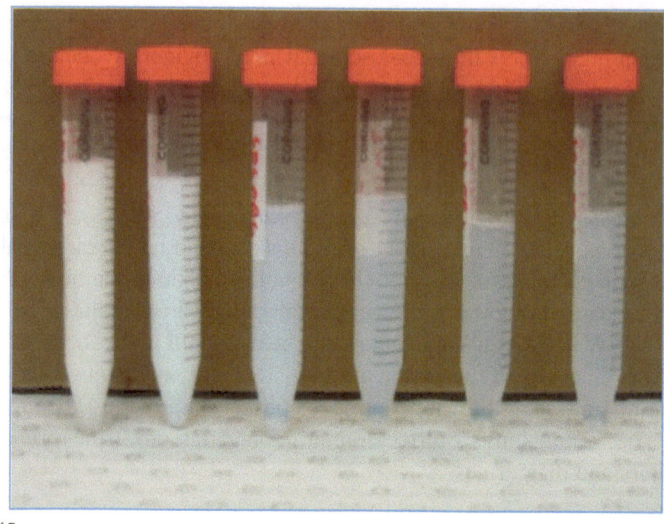

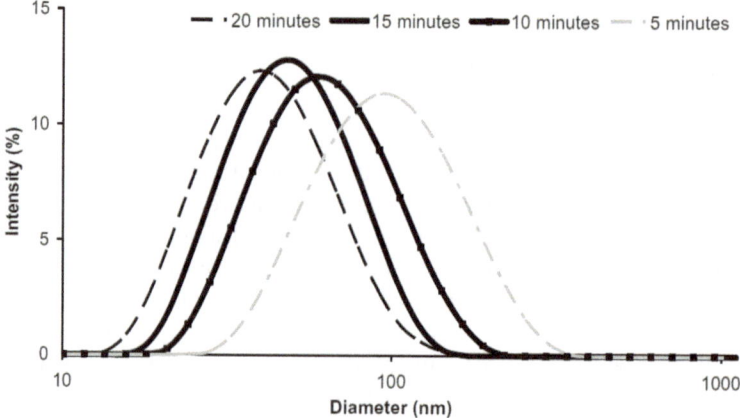

Fig. 1.3 a Visual appearance of sample emulsions created using 15 wt% sunflower oil with 20 kHz ultrasound (50% amplitude) varied from 5 to 20 min. **b** The effect of increasing sonication duration towards the emulsion droplet size distribution. Reprinted from Leong et al. [8], Copyright 2009, with permission from Elsevier

an emulsion with smaller droplets will tend to have a higher viscosity [10, 11]. Studies have correlated the 'creaminess' properties of an emulsion, and in terms of perceivable fat content, it was found that this generally increases with the viscosity of the emulsion and the concentration of fat [11].

1.2   Emulsion Creation

The creation of an emulsion requires an input of energy to create the physical shear forces needed to disperse one phase into the other. Mechanical work or a chemically favourable driving force, such as phase inversion, is required to generate small emulsion droplets. In order to overcome the natural tendency for the immiscible phases to separate, an emulsifier loaded into the system will position itself at the newly formed droplet interfaces, preventing spontaneous coalescence.

In most industrial processes, the dispersed phase is emulsified into the continuous phase by means of high energy shear, mixing and turbulence [12]. Some commonly used techniques to form emulsions include rotor–stator type devices, high-pressure homogenization (HPH), microfluidization (MF) and ultrasonication (US). US is an emulsification method that possesses several advantages in regard to energy efficiency, droplet size reduction and ease of operation compared to other known techniques [6].

1.3   Ultrasound

Sound waves are pressure waves that oscillate at a characteristic frequency as they propagate through a medium. Ultrasound refers to sound waves that oscillate at a frequency beyond the limits of human hearing. The frequency of ultrasound that is used to perform processing and sonochemistry generally spans the frequency range between 16 and 3000 kHz.

Although not audible when transmitted through air, ultrasonic waves transmitted through a liquid will induce a phenomenon known as acoustic cavitation [13]. Acoustic cavitation is the nucleation, growth and collapse of gaseous bubbles within a fluid [14] (Fig. 1.4). The nucleation event typically involves dissolved gas bubbles already present in the fluid, but can also be initiated from microscopic pockets of gases trapped in crevices or on surfaces of motes/solid materials dispersed within a fluid. Bubbles that are exposed to ultrasonic frequencies between 20 and 100 kHz at a high acoustic amplitude can undergo intense growth and collapse, known as transient cavitation. Under these conditions, bubbles that reach a size within what is known as the resonance size range will expand dramatically during the negative cycle of the pressure wave (which can be up to approximately 10–20 times the initial radius [15]), before imploding strongly during the positive pressure cycle. This collapse can be an extremely high energy event, leading to the production of extreme localized temperatures within the core of the collapsing bubble, and creation of powerful shear forces. Collapsing bubbles may reach temperatures of many thousands of degrees Celsius, but these extreme temperatures are confined to small areas at the core of the collapsing bubble and near the bubble surface [16].

The shear forces manifest primarily in the form of pressure shock waves, liquid microjets and acoustic streaming. Pressure shock waves caused by bubble collapse

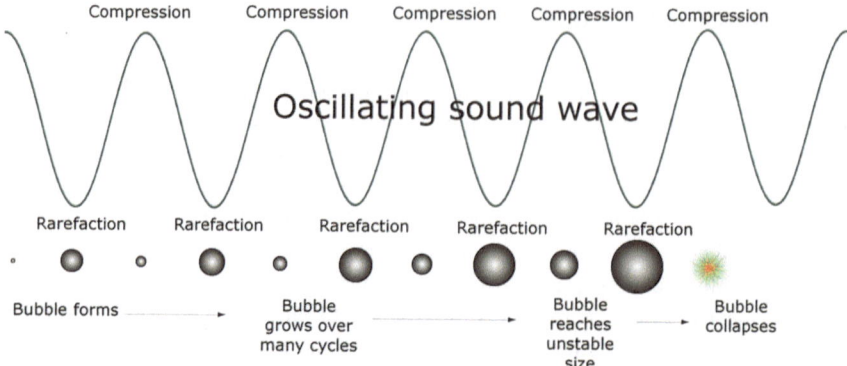

Fig. 1.4 Schematic depiction of bubble nucleation, growth and collapse under the influence of an oscillating sound wave. Reprinted with permission from Leong et al. [14]

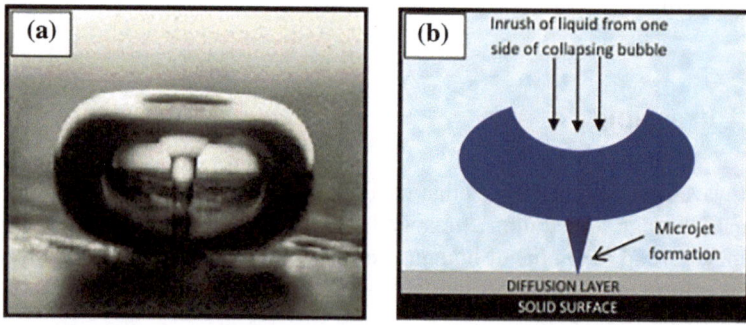

Fig. 1.5 Microjet formation from a collapsing bubble occurring near a solid surface. Sourced from the open access reference by Collyer et al. [20]

can reach pressures up to several hundred atmospheres. These shock waves form due to a symmetrical bubble collapse and then propagate radially outwards from the collapse point into the surrounding fluid. Often accompanying these shock waves is an effect known as acoustic streaming [17], which can be observed as rapid fluid flow induced by the oscillation of bubbles. Acoustic streaming occurs due to attenuation of sound energy within the medium.

If the bubble collapse occurs near a surface, it is usually of an asymmetric nature. The collapse may lead to unidirectional expulsion of high-velocity jets into the surrounding fluid, known as microjetting (Fig. 1.5). Naude and Ellis [18] hypothesized that microjets formed by collapsing bubbles were the cause of observed pitting of solid surfaces and particle size reduction of colloids on exposure to ultrasound. These microjets have velocities in the order of 100 m/s [19] and, together with shock waves and acoustic streaming, can facilitate extremely rapid bulk mixing in fluids that can enhance mass transfer across interfacial boundaries in multiphase systems.

The strong shear forces generated by acoustic cavitation are extremely effective at producing emulsions. Indeed, one of the first known applications of ultrasound in fluid processing, reported in 1927, was for the purpose of emulsification [21]. Temperature hot spots created by collapsing bubbles [22] may also facilitate functional changes in heat-sensitive components. In systems that contain proteins (e.g. milk), these proteins can become partially denatured by the localized heating, improving their surface activity.

1.4 Mechanisms of Ultrasonic Emulsification

Ultrasonic emulsification occurs via two main processes [7]. First, the phase to be dispersed is erupted into the continuous phase as large droplets by interfacial waves caused by the propagation of ultrasound. Second, the intense physical shearing effects generated within the continuous phase during acoustic cavitation cause the gradual breakdown of the initially formed large droplets. With continued sonication, size reduction proceeds until a size limit is reached. This limit is dependent on several properties including the amount of emulsifier, the rate of droplet coalescence caused by inter-droplet collisions during processing, the fluid viscosity and operating temperature. Droplet–droplet coalescence can be minimized by ensuring that fast stabilization of new interfaces occurs by using a sufficient amount of effective emulsifier.

1.5 Emulsification Set-up and Conditions

A common ultrasonic emulsification set-up consists of a horn-type sonotrode with its active surface submerged into a container holding the mixture of immiscible phases and an emulsifier. While ultrasonic disrupter horns are capable of delivering very high-intensity ultrasonic waves (typically ranging between 10 and 1000 W/cm^2) into a fluid, the active region is typically confined to a relatively small area with an effective distance in the order of several *cm* from the tip surface. To maximize the effectiveness of the acoustic cavitation shear forces generated, the holding container should have a cross-sectional area that is not excessively larger than the horn tip, and the volume being processed should be small enough to ensure uniform treatment occurs within the desired processing time. Larger horn tips are available that are capable of processing larger volumes of material, but it is generally recognized that scale-up of ultrasonic systems can be quite challenging due to the fact that the emulsification activity is localized near the sonotrode surface.

For a batch ultrasonic emulsification process, if the immiscible phases are initially two separate continuous layers, it is recommended that the horn tip be positioned at or near where the two interfaces meet (Fig. 1.6a). The interfacial waves will immediately disperse the one phase into the other as large droplets. Once this occurs, ongoing size

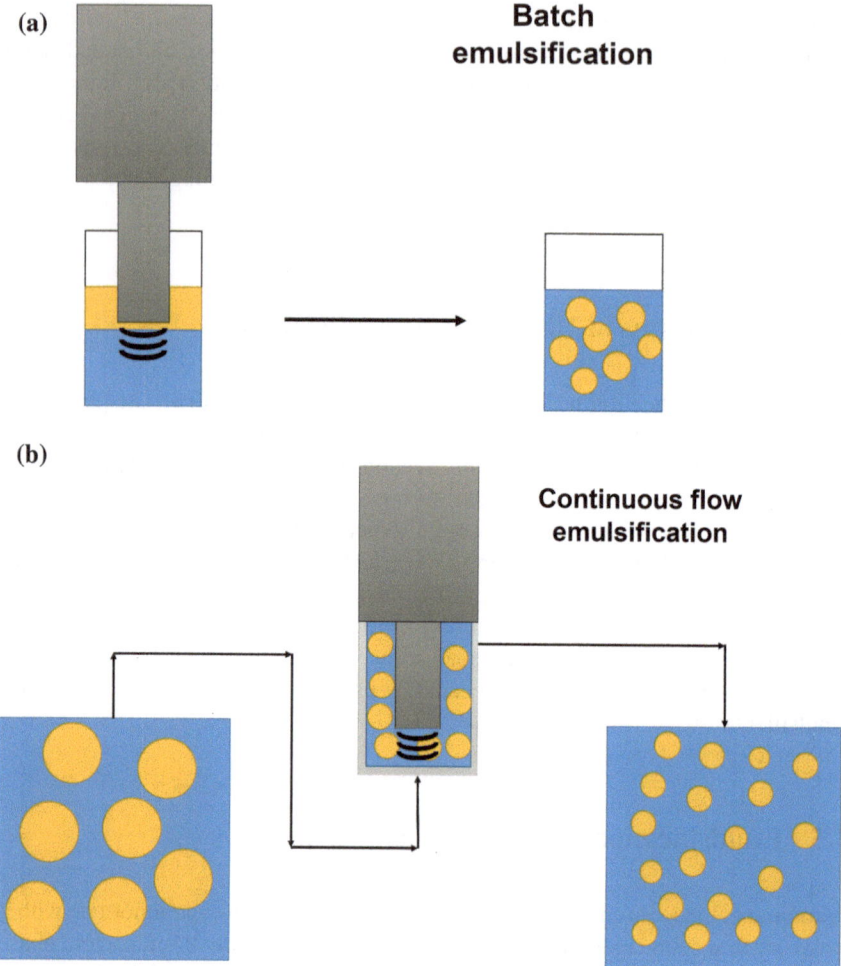

Fig. 1.6 Schematic depiction of a typical **a** batch and **b** continuous flow ultrasonic emulsification set-up using an ultrasonic disruptor horn

reduction will proceed due to acoustic cavitation. In small volume batches, it is not necessary to premix the two phases together [7], and sonication will generate quite reproducible outcomes provided a sufficient processing time is employed. Generally, a batch production of 30–75 mL can be homogenously emulsified without premixing by using ultrasonic power of ~30 to 50 W (calorimetric energy delivered by an 11 mm horn) for a minimum of 2–5 min.

As mentioned, scale-up using larger horns and larger containers is not a straight-forward process. Due to limitations in regard to the active region of processing, it is difficult to ensure homogeneous treatment of larger volumes of material using a batch process. It is better to process large volumes using a continuous flow-through unit that can ensure all fluid elements are exposed to the active sonication region. Continuous flow-through cells are available that attach directly to a disruptor horn (Fig. 1.6b). These cells enable optimal delivery of ultrasonic energy to fluid that is continually flowed through. In this way, a large volume of fluid flowed through the cell can be subject to a uniform treatment by the applied ultrasound. Multiple flow-through units can be operated in parallel to increase the throughput or positioned in series to increase the effective residence time that will enable the production of emulsions with smaller EDS.

1.6 Comparison of Ultrasonic Emulsification with Other High Shear Emulsification Methods

Industrial production of sub-micron emulsions is based on the use of high shear to disrupt and distribute small droplets of the dispersed phase throughout the continuous phase. Ultrasonic emulsification has been compared with most conventional and state-of-the-art emulsion preparation techniques. Some common methods used in industrial processes are rotor–stator (high-speed mixers), high-pressure homogenizers and microfluidizers [23]. In these techniques, the disruption and dispersion of droplets of one phase into the other are achieved by the generation of intense shearing forces.

Rotor–stator devices create shear by high-speed rotation of an impellor within a static enclosure. The rotor–stator elements are designed with channels that can maximize the shear forces generated for a given rate of revolution. Compared to the other techniques, the shear forces they produce are relatively low, meaning they are not able to create emulsions with very small EDS. Nonetheless, they are commonly used for the preparation of microemulsions, for instance the production of mayonnaise and sauces.

Microfluidizer®, a proprietary technology offered by Microfluidics™, is often recognized as the most effective nanoemulsion preparation technique presently available [12], creating emulsions with good energy efficiency. A Microfluidizer operates by dividing a stream flowing under high pressure in two and then redirecting them into a central flow chamber where they collide. This collision produces intense shear and impact that reduce the emulsion droplet sizes. Due to the potentially high capital and maintenance costs involved, these devices are typically used for applications in the pharmaceutical industry for creating higher valued emulsion products.

A study performed by Jafari et al. [12] compared emulsion preparation using ultrasound at matched energy input with a Microfluidizer. Comparable performance in terms of emulsion size reduction was observed when using matched 20 kJ/kg energy input. The Microfluidizer achieved a mean volume-weighted particle size of 0.83 μm compared with 1.02 μm for ultrasonication at 20 kHz. Although ultrasound was applied in a batch reactor for this comparison (~400 mL), larger volumes of fluid can be processed by using continuous flow-type arrangements (Fig. 1.6b). Scale-up of ultrasonic processing can be achieved by providing the equivalent energy input by controlling the residence time of fluid within the reactor. The time dependence of ultrasonication emulsification means that emulsions can be produced to a desired droplet size.

High-pressure homogenization involves passing fluids through a narrow orifice driven by a high-pressure drop across a valve. High-pressure homogenizers are widely used as industrial-scale emulsification units, for instance for the homogenization of milk, and are one of the most effective methods by which nanoemulsions can be formed. In these devices, the sudden restriction of flow under high pressure creates extreme turbulence, cavitation, high shear and inertial forces [24]. Pressures up to 2500 bar are readily achievable, enabling the disruption of the dispersed phase into very fine droplets and the creation of emulsions with a mean droplet diameter of less than 0.2 μm [25]. Some drawbacks of high-pressure homogenizers include high operating costs due to energy consumption and maintenance.

One of the mechanisms for emulsification in ultrasonication and high-pressure homogenization is the creation of cavitation bubbles that, upon collapse, generates extreme shear and pressure shock waves. For this reason, ultrasonication and homogenization techniques are clearly superior to high-speed mixing methods [26]. For a given energy input, ultrasonic emulsification is capable of achieving comparable but slightly less effective size reduction compared with high-pressure homogenizer or Microfluidizer systems [12].

With homogenizer systems, all of the fluid passes through the narrow valve or orifice, helping to ensure even processing and uniform droplet size distributions. In ultrasonic emulsification, the shear forces are generated by the acoustic cavitation bubbles which are generally confined to a small region, and so the uniformity of processing is generally lower compared with the aforementioned homogenizer methods. Depending on the volume being processed, an extended processing time is required to enable the entire fluid to be subject to the effects of the cavitation bubbles. As mentioned, a continuous flow application (Fig. 1.6b) can be employed to maximize the delivery of ultrasonic energy into large volumes of fluid and provide more uniform treatment. One advantage ultrasonication has over homogenization techniques is its ease in terms of cleaning and maintenance as there are no moving parts and fewer narrow flow areas that could be clogged or damaged.

1.7 Theoretical Understanding of Emulsion Formation

The process of emulsification requires energy, and it is important to understand the basic interactions that lead to emulsion formation in order to design processes that are able to maximize energy efficiency. The physical shear and turbulence generated in ultrasonic systems originate from the phenomenon of acoustic cavitation. The cumulative effect of thousands of these miniature implosions forms the basis of ultrasonic emulsification [6, 27–29]. The intensity of acoustic cavitation induced in fluids from the application of low-frequency ultrasound generates several physical effects important to emulsification, namely microstreaming, microjetting and shock waves [30–32]. The high turbulence and velocity gradients generated over very small length scales that arise from these effects lead to the production of nanoemulsion [34, 35].

The basic theory regarding emulsification in turbulent flow was established half a century ago. To date, an extensive understanding of the theory has been established to predict droplet size formation based on the applied energy [35–39]. Two regimes of emulsification were distinguished, these being the *turbulent inertial* and *turbulent viscous* regimes [39]. The difference between the two regimes is the relative size of the droplets, d, to the size of the smallest eddies generated in the fluid flow, λ_0. The size of the smallest eddies in the turbulent flow, λ_0, which is also referred to as the 'Kolmogorov scale' can be defined according to the following empirical relationship that reflects the hydrodynamic conditions during emulsification:

$$\lambda_0 \approx \varepsilon^{-1/4} \eta_c^{-3/4} \rho_c^{-3/4}, \tag{1.1}$$

where η_c is the viscosity ($m^2 s^{-1}$) and ρ_c is the mass density of the continuous phase ($kg\ m^{-3}$), while ε is the average power dissipated per unit mass of the fluid (W kg^{-1}). In the turbulent inertial regime, the relative droplet sizes formed are larger than the smallest eddies, $d > \lambda_0$, whereas in the turbulent viscous regime, the droplet sizes are smaller than these eddies, $d < \lambda_0$.

During acoustic cavitation, microstreaming, microjetting and microturbulence are the main sources of shear stress that generate the turbulent eddies. The scale of these actions can be assumed to be of a similar scale as the resonance size of a cavitation bubble, which is typically 2 μm [32]. For a US system, assuming λ_0 as 2 μm, the energy dissipation rate can be calculated to be in the order of 6×10^4 W/kg.

Another important consideration is the critical drop diameter, d_{crit}. This is the average droplet size that exists once steady state has been reached, in which the rate of droplet break-up and recoalescence are in balance. In the turbulent inertial regime, d_{crit} is expressed as follows [39]:

$$d_{crit} = C\varepsilon^{-2/5} \sigma^{3/5} \rho_C^{-3/5}, \tag{1.2}$$

where C is a constant of proportionality of the order of unity, ε is the power density (i.e. the average power dissipated per unit mass), σ is interfacial tension and ρ_C is the density of the continuous phase. Note that in this correlation it is assumed that the contribution of the droplet phase viscosity is negligible. This diameter can also be replaced with a value that represents the volume mean, Sauter mean or Z-average diameter, in which case the correlation generally holds with a slight modification of the constant [26, 33].

Equation (1.2) provides a good prediction of droplet size produced based on power density, and it can be generalized that $d \sim \varepsilon^{-0.4}$ for the formation of macroemulsions (0.1–5 μm). This basic correlation can be widely applied to compare the relative energy efficiencies between different processing techniques such as ultrasonication (US), ULTRA-TURRAX (UT) and high-pressure homogenization (HPH) systems

[26]. One notable feature of this equation is that it does not allow for determination of the emulsification kinetics, due to a lack of time dependence.

To account for time dependence, another commonly reported quantity is the energy density (E_v), which is the energy input per unit volume, per unit time [33, 40]:

$$E_v = \frac{P\tau}{V} = \frac{P}{v},$$ (1.3)

where P is the power input (W), τ is the residence time (s), V is the processing volume (mL) and v is the flow rate (mL s^{-1}).

Chapter 2
Selection of Operating Parameters

Abstract There are a number of parameters that influence the effectiveness of ultrasonic emulsification and these should be considered when operating or developing an ultrasonic emulsification protocol. This Chapter provides a brief guide to selecting appropriate operating conditions.

Keywords Power · Gas content · Pressure · Temperature · Emulsifier
Viscosity · Ultrasound frequency · Acoustic shielding

2.1 Influence of Power and Processing Time

The power delivered to an ultrasonic transducer influences the amplitude of the ultrasonic waves. Stronger amplitude ultrasound will result in more intense acoustic cavitation effects and hence more effective emulsification per unit time.

In general, a higher power delivered will increase the rate at which emulsion droplets are broken up into smaller droplets. While increasing the power will increase the rate of emulsion droplet size reduction, there exists a limit to which increasing power will improve the effectiveness of size reduction due to a phenomena known as 'acoustic shielding' that will counter further benefit. The cavitation bubbles that are nucleated in the system will tend to accumulate at the pressure anti-nodal regions due to the influence of acoustic forces known as Bjerknes forces. As they accumulate, the bubbles will form into clusters or 'clouds'. Bubbles that are situated near the centre of these clouds will experience a lower effective pressure compared with the bubbles nearer the surface (Fig. 2.1), creating a 'shielding' effect [41]. In a given system, there exists an 'optimal' power/amplitude for which maximal cavitation and hence emulsification can be achieved efficiently to avoid the effects of acoustic shielding. This is one of the reasons as to why scale-up of ultrasonic systems can be a challenge, since there is a limit to how much the power can be increased before the effectiveness of cavitation decreases.

© The Author(s), under exclusive licence to Springer International
Publishing AG, part of Springer Nature 2018
T. S. H. Leong et al., *Ultrasonic Production of Nano-emulsions for Bioactive Delivery in Drug and Food Applications*, Ultrasound and Sonochemistry, https://doi.org/10.1007/978-3-319-73491-0_2

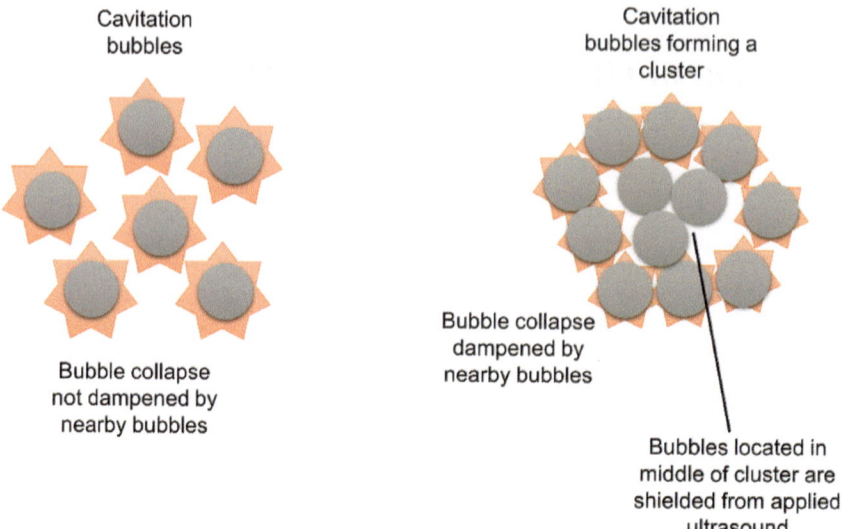

Fig. 2.1 Schematic depiction of cavitation bubbles forming a cluster that results in acoustic shielding effects

Another factor that can limit the amount of power that should be applied is known as 'over-processing' [6]. The high shear forces that enable the disruption of oil droplets also increase the chance of droplets colliding together to reform into large droplets. Higher processing power gives the droplets greater momentum, increasing the probability of coalescence upon collision. Over-processing is a problem that also exists in other types of emulsification techniques such as high-pressure homogenization and microfluidization. Over-processing can be avoided or minimized by ensuring that there is sufficient emulsifier present in the system to rapidly stabilize interfaces and prevent coalescence occurring despite the more powerful collisions.

It should be noted that the electrical power delivered to an ultrasonic transducer is not all transformed into usable work, although the electrical power drawn for a given system is usually proportional to the effective power delivered to the fluid. A well-accepted method, by which the delivered power can be estimated, is by determining the amount of energy that becomes transformed into heat. This is known as the calorimetric power of the system [42] and accounts for the heat released by acoustic cavitation during bubble collapse, as well as frictional heating due to intense fluid motion and heating loss from the transducer itself. Typically, to form nanosized droplets will require an energy input (as determined by calorimetry) in the order of 20–1000 kJ/kg of material depending on the amount and type of surfactant, the viscosity of the fluids, and the dispersed phase volume.

2.2   Influence of Gas Content

One of the prerequisites for efficient acoustic cavitation is the presence of dissolved gas nuclei in the fluid being treated. While the actual amount of gas present in the system only has a slight influence on the emulsification effectiveness [40], a fully degassed or over-gassed system may affect the degree of cavitation produced.

A degassed system will have fewer nuclei to undergo acoustic cavitation growth and collapse. If driven at a sufficiently high power, acoustic cavitation will still be produced in a fully degassed system due to nucleation sites that exist on crevices of surfaces and particles. Interestingly, bubbles in an under-gassed system will tend to undergo more intense collapse, as there are fewer bubbles present in the system to cushion and absorb the energy released. It has been observed that a single cavitation bubble not influenced by other bubbles, reaches a higher core temperature during collapse than for interacting bubbles [16].

By contrast, in highly gas saturated systems, nucleation and cavitation occur readily. However, the elevated gas content in the system leads to an increase in the gas/vapour pressure ratio inside the bubbles. The consequence is that bubble collapse events become cushioned, reducing the intensity of shock waves and other shear forces responsible for emulsification, in a similar manner to acoustic shielding.

2.3   Influence of Ambient Pressure

The extent of pressurization of a system will also play a role in the efficiency of the acoustic cavitation produced and the emulsification outcome. This overpressure (pressure above ambient conditions) has a critical role in determining the parameter thresholds for which cavitation occurs in a given system. For a given bubble size, an elevated static pressure will increase the thresholds (pressure/radius) at which it will undergo acoustic cavitation. Bubbles with higher threshold criteria for acoustic cavitation will have a lower probability of undergoing cavitation. Although there are fewer bubbles undergoing acoustic cavitation with increasing static pressure, the accompanying shock waves produced by bubbles that do collapse at elevated ambient pressure are more intense.

Studies have found that there is usually an optimal overpressure for improving emulsification efficiency that is a compromise between the increased shock wave intensities produced from collapsing bubbles, and the decreased number of bubbles undergoing cavitation. Bondy and Sollner [43] suggested a pressure of approximately 200 kPa above ambient was optimal, while Leong et al. [8] reported an optimal overpressure of approximately 300 kPa (Fig. 2.2). However, in the study by Leong et al. [8], application of overpressures greater than 400 kPa resulted in ineffective cavitation, and no emulsification.

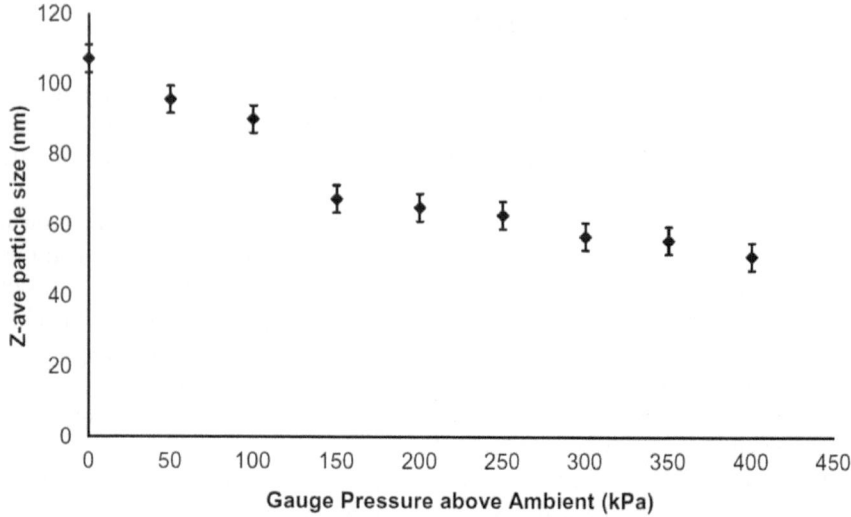

Fig. 2.2 Emulsion droplet size as a function of applied pressure above ambient for a fixed sonication time and amplitude setting. Note that no emulsification was achievable past 450 kPa pressure. Reprinted from Leong et al. [8], Copyright 2009, with permission from Elsevier

2.4 Influence of Temperature

The temperature of the system influences several underlying properties of the emulsion including its viscosity, interfacial properties and gas content. The influence of temperature on emulsification is complex. In general, temperatures between ~15 and 50 °C have no significant effect on emulsification effectiveness.

Elevated temperatures reduce fluid viscosity and tend to increase the number of nuclei available in the liquid, making droplet break-up more efficient. Countering this, as the solution temperature rises the vapour pressure increases (i.e. the pressure inside the bubble). A higher vapour pressure tends to cushion the bubble collapse, thereby reducing the collapse intensity [29].

In addition, high temperatures may affect the effective emulsifying properties of surface active agents. For example, proteins which are natural emulsifiers may denature at critically high temperatures. This may increase their effectiveness by exposing hydrophobic residues, but may render them ineffective if they form large aggregates. Elevated temperatures may also instigate more foaming, which can promote phase separation. Due to these reasons, an effective temperature control method is usually used during ultrasonic emulsification to prevent excessive temperature rise. Marie et al. [44] found that during high-pressure jet emulsification (a cavitation-based emulsification method similar in principle to microfluidization and high-pressure homogenization) smaller and more monodispersed droplets could be formed with cooling. In this case, a 39% reduction in droplet diameter was achieved compared with the absence of cooling.

2.5 Emulsifier Selection and Concentration

Having a sufficient concentration of emulsifier (i.e. surfactant) present in the mixture to stabilize any newly formed interfaces is critical for effective emulsion creation [29]. Without sufficient amounts of emulsifier, phase separation will occur rapidly as droplets recombine to minimize their surface-free energy. The stabilization of formed emulsion droplets is largely due to the creation of repulsive forces (electrostatic and/or steric) on the surface as they become coated with surfactant molecules.

In general, increasing the amount of emulsifier in the system will lead to more effective emulsion formation. Increasing amounts of surfactant will increase the amount of interfacial area that can be stabilized, enabling the formation of smaller droplets for a given volume fraction of the dispersed phase. If foaming occurs during emulsification, some of the emulsifier goes to the air–water interface of the foam, reducing the amount of surfactant available to stabilize the emulsion droplets. For this reason, the amount of surfactant required for ultrasound emulsification can be lower than for other techniques, such as rotor–stator devices, which can generate a lot of foam due to vigorous agitation.

The type of emulsifier used is an important consideration. First and foremost, the emulsifier needs to be nontoxic and be suitable for consumption or topical application for foods and pharmaceuticals. Secondly, it should have good surface activity that will enable rapid interfacing at newly formed surfaces and be able to reduce the surface/interfacial tension of the two phases effectively. A lowering of the surface tension makes it relatively easier for applied shear to overcome the surface energy in order to disperse the immiscible liquids into one another. The type of head group and the length of the hydrocarbon chain are also key determinants of the emulsifier properties.

Low molecular weight surfactants are highly mobile surfactants that are able to rapidly interface at immiscible phase boundaries [1]. Examples of these types of surfactants include fatty alcohols, glycolipids and fatty acids, all of which can be derived from oils or synthesized from other starting materials. High molecular weight surfactants include proteins, polysaccharides and other polymeric-type surfactants [1]. For food and pharmaceutical applications, these emulsifiers tend to be derived from naturally occurring sources such as milk proteins, hydrocolloid gums and lecithins. Natural surfactants are of particular interest in food and drug applications due to their low toxicity and their ability to provide nutritional benefits in addition to their surface active functionality. Examples [45, 46] of the use of ultrasonication to emulsify oils into skim milk proteins will be presented later in Sect. 3.2.1.

Another consideration is the hydrophilic/lipophilic balance (HLB) of the surfactant, which is a measure of how hydrophilic or lipophilic the emulsifier is. The HLB determines whether or not the resulting emulsion formed will be an O/W or W/O emulsion. Low HLB emulsifiers (HLB = 1 − 4) favour the formation of W/O emulsions, while higher HLB emulsifiers (7–12) favour the formation of O/W emulsions [1]. Some common classes of surfactants used for food processing with their HLB values are presented in Table 2.1. When forming multiple/double emulsions, a com-

Table 2.1 Common classes of emulsifiers used in food processing ranked in increasing HLB strength

Emulsifier	HLB
Saturated and unsaturated mono/diglycerides	3–4
Propylene glycol esters	3.5
Sorbitan monostearate	4.7
Lecithin	4–9
Polyglycerol esters of fatty acids	7
Cellulose gums	10–11
Polysorbate 65	11.0
Sucrose esters	11–15
Polysorbate 60	14.5
Polysorbate 80	15.0
Sodium stearoyl lactylate	21.0

HLB values sourced from [47]

bination of low and high HLB emulsifiers needs to be employed in the inner and external phase to stabilize their respective dispersed phases. The ratio of the amount of surfactant used needs to be carefully balanced to avoid competitive transport of internalized disperse phase to the external phase or vice versa.

2.6 Influence of Continuous Phase Viscosity

The viscosity of the continuous phase influences the overall stability of the dispersed phase. Increasing the continuous phase viscosity reduces the mobility of the dispersed droplets, slowing down inter-droplet collisions that may lead to droplet coalescence, and slows the rate of creaming. For some systems, it can be purposefully elevated to improve EDS reduction efficiency [23] and long-term stability. This is usually achieved by the addition of agents known as 'stabilizers' [23]. Stabilizers should not be confused with emulsifiers. Although these macromolecules are surface active, they are not used to provide direct inter-droplet emulsion stability through steric/electrostatic effects, but added as a co-surfactant to increase the viscosity of the continuous phase. Commonly used stabilizers suitable for food-grade emulsions include polyethylene glycol (PEG) and hydrocolloid gums such as xanthan gum, locust bean gum and carrageenan.

It should be noted that the viscosity of the continuous phase will affect the onset of acoustic cavitation, as the formation of 'cavities', i.e. nucleation of bubbles from ultrasonication, becomes more difficult with higher viscosity. Bubble collapse in high viscosity fluids does however tend to be more intense due to enhanced inertia of the bubble motion [48].

2.7 Ultrasonic Frequency

Ultrasonic emulsification is typically achieved using low-frequency ultrasound (i.e. 20–100 kHz). The cavitation bubbles formed in this frequency range tend to undergo strong transient cavitation collapse, which produces strong shear forces that will disrupt and break apart emulsion droplets into smaller droplets. Horn-type sonotrodes typically deliver ultrasound at a fixed frequency of ~20 kHz, while bath-type and plate-type ultrasonic equipment can be tuned to provide slightly higher frequency ultrasound. Horn-type sonotrodes are more commonly used as they are more effective for creation of nanosized emulsion droplets owing to the high-intensity ultrasound delivered in a narrow cavitation zone. Bath-type ultrasound equipment can be effective in delivering ultrasound across a wider region and is useful for emulsions requiring less-intensive size disruption.

Higher frequency ultrasound in the MHz range can be used to assist in the formation of nanosized emulsion droplets through a process known as *tandem acoustic emulsification* [49, 50]. Nakabayashi et al. [49] used a sequentially increasing ultrasonic frequency protocol (20 kHz–1.6 MHz and then finally 2.4 MHz) to produce highly stable nanoemulsions of an ethylenedioxythiophene monomer. The emulsions formed in this process were stable for 1–2 years and, due to the small and monodispersed size of the droplets formed (<100 nm), they were stable over this period without requiring the use of any emulsifier.

Mechanistically, droplets are initially formed by the application of 20 kHz ultrasound. The subsequent exposure of the formed emulsion to higher frequency ultrasound in the MHz range causes the droplets and the surrounding solvent to experience dramatic acceleration (due to the development of acoustic radiation forces and acoustic streaming) [50]. This causes the droplets to collide together and break apart further into smaller droplets (the collision forces are high enough to induce break-up rather than coalescence). The acceleration becomes stronger with increased frequency, which is why the order of the applied frequency is important for tandem acoustic emulsification.

2.8 Issues to Consider

2.8.1 Effect of Ultrasonication on Product Quality

The effects of cavitation bubble collapse lead to an increase in temperature, pressure and the generation of chemical radicals in the form of hydrogen and hydroxide radicals [51]. These effects may compromise the quality of the emulsion formed, namely by degradation of the emulsifier and/or the organic phase. In general, it has been shown that the degradation is usually slow with respect to the rate of emulsification, such that product quality is not diminished except for instances in which very long sonication times are required [29]. The oxidation of lipids is linked to the formation of free radicals. The production of radicals occurs much more efficiently at mid-range ultrasonic frequencies (i.e. between 400 and 800 kHz) [52, 53] and tends to be minimal at 20 kHz [51]. In milk-based emulsions, Juliano et al. [54] detected formation of compounds due to oxidation of milk above human sensory thresholds when using long durations of treatment at high power (specific energy >300 kJ/kg).

Where possible, product quality loss due to ultrasonication should be controlled by employing the shortest possible contact time/low energy density for desired emulsion specifications [55]. However, in some cases the high energy required (up to 1000 kJ/kg) to produce an emulsion with nanosized droplets (e.g. <50 nm) means that long contact times together with high energy density, and therefore some product degradation, are unavoidable. In these instances, the food emulsion should be designed to use oils that are more resistant to oxidative degradation.

The ultrasonication of oils with a large proportion of poly-unsaturated fatty acids, e.g. sunflower oil [56], can be particularly prone to oxidative degradation by applied ultrasound (low-frequency, high-power ultrasound). A study by Chemat et al. [56] showed a significant increase in the production of hexanal after ultrasonication, which gives a pungent, grassy odour. Extended ultrasonication of oils can also result in a burnt metallic-like odour in the product. This odour can be attributed to the formation of (Z)-hept-2-enal and (2E,4E)-deca-2,4-dienal volatiles, which contribute a fishy and deep-fried odour, respectively [56]. Note that the degree of oxidation is oil dependent. Peanut and olive oils, which contain less poly-unsaturated fatty acids, are less prone to damage by the effects of ultrasound.

The shearing energy produced will generally not result in physical breakage of covalent chemical bonds, and so emulsifiers present in the system are usually not affected. However, natural surfactants such as proteins have secondary, tertiary and quaternary structures, which may be affected by ultrasound. For example in milk systems, ultrasound has been shown to affect the surface hydrophobicity of the proteins present [55, 57, 58]. This partial denaturation of proteins present in milk can be used to improve the interfacial stabilization of emulsion droplet interfaces (see Chap. 3).

2.8.2   Formation of Metal Particulates

The forces generated by acoustic cavitation will result in wear and tear of metallic surfaces within the active area of emulsification. Horn-type ultrasonic probes operating at low frequency (20–40 kHz) and high power (10–1000 W/cm^2) will result in gradual erosion of the horn tip or container surfaces. As such, there is a possibility that ultrasonic treatment could lead to formation and release of metallic particles directly into the processed food material. These metallic particles could contaminate the food product and be very difficult to remove, particularly if they are nanosized.

Recently, a study by Mawson et al. [59] found no evidence of production of metallic nanoparticles (<80 nm) that are considered to be particularly deleterious to health during sonication using 20 kHz ultrasound. Nanopore-sized filters were used to selectively remove any formed particulates from the treated fluids, which were then imaged using field emission scanning electron microscopy. In their study, no nanoparticulate material was observed even for prolonged operation periods of up to 7.5 h ultrasound exposure using a 20 kHz ultrasonic horn. These long processing times are typically not required in food processing applications, and studies [51] have shown that the useful effects of ultrasound are usually achieved within the first few seconds of application. The outcomes from Mawson et al. confirm that, within these limits, ultrasonication can be safely applied to food products. For emulsification, the initial dispersion occurs within the first few seconds of ultrasound application, with further EDS reduction by droplet break-up continuing with increasing duration of application (generally up to several minutes). As noted above, excessive contact time should be avoided to prevent degradation of oxidation-sensitive materials.

A strategy to avoid metallic release into foods is to use metal-free containers and surfaces. For example, Freitas et al. [60] used a steel jacket to transmit sound waves via pressurized water to a glass tube that was installed inside the jacket. In this way, no metallic particles were emitted into the sonicated fluid.

With extended usage, the tip of an ultrasonic horn will eventually become eroded due to strong cavitation effects. The tip can be periodically replaced relatively cheaply, especially compared with replacement or repair of flow chambers such as those used in microfluidizers or high-pressure homogenizers.

Chapter 3
Applications of Ultrasonic Emulsification

Abstract A large number of studies have demonstrated the efficacy of ultrasonic emulsification for a growing range of applications. A selection of studies using ultra-sonication to emulsify oils for different functions is highlighted in Table 3.1. Selected applications for foods and pharmaceuticals will be showcased in this chapter. While the formation of nanoemulsions are not exclusive to ultrasonics, the acoustic cavitation mechanism used has been shown to provide additional benefits that improve emulsion stability and/or provide additional functionality to the resultant emulsion.

Keywords Functional foods · Pharmaceuticals · Proteins · Dairy · Essential oils Anticancer · Drug delivery

3.1 Applications in Pharmaceuticals

3.1.1 Delivery of Anticancer Drugs

Emerging nanotechnology has solutions to the problems existing with conventional pharmaceutically active ingredients or the dosage forms which incorporate these components. In this context, nanoemulsions act as an effective carrier for a range of components that are difficult to solubilize in water [61, 62]. Pharmaceutical nanoemulsions are isotropic and kinetically stable drug delivery systems where the diameter of droplets is usually less than 500 nm. They may be transparent, translucent or milky in nature. These nanoemulsions are generated either as oil-in-water (O/W) or as water-in-oil (W/O) systems. The major advantage of these carrier systems over conventional delivery systems, [63] such as simple gel/polymer-coated liquid tablets, are that they increase the rate of dissolution of the drug through the digestive system, and hence absorption across the gastrointestinal (GI) tract and into the bloodstream of the patient, thereby enabling increased bioavailability and concentration at the targeted active site, e.g. tumour cells. An increase in the bioavailability could be due

T. S. H. Leong et al., *Ultrasonic Production of Nano-emulsions for Bioactive Delivery in Drug and Food Applications*, Ultrasound and Sonochemistry, https://doi.org/10.1007/978-3-319-73491-0_3

to a range of reasons including an improvement in the solubilization of drug, protection against enzymatic hydrolysis, an increase in the surface area of droplets which lead to wider distribution within the body, an easier absorption in the GI tract and surfactant-induced permeability changes of the drug across cell membranes. More importantly, the droplet size plays an important role as it relates to the dissolution of drug components into the digestive system. In general, nanoemulsions have appeal as drug delivery systems as they are efficient, convenient, flexible and more patient compliant [3, 61].

As discussed above, conventional technologies used to generate these pharmaceutical nanoemulsions have various disadvantages such as requiring high amounts of stabilizers or surfactants, higher amounts of energy and potential instability issues. Ultrasound can be relatively energy-efficient for producing homogeneous emulsion systems and provides flexibility in controlling the particle/droplet size to confer long-term stability. Ultrasonic emulsification appears to be competitive or even superior in terms of producing emulsions with a small and uniform droplet size and has energy efficiency comparable to other conventional systems such as high-pressure homogenizers. Thus, ultrasound is a capable, versatile and robust technique for generating a range of pharmaceutical nanoemulsions, which can act as new nanocarriers as improved drug delivery vehicles. The importance of ultrasound as a technique for the generation of nanoemulsions has been reviewed recently by Sivakumar et al. [64].

Curcumin, a yellow polyphenolic phytochemical obtained from Curcuma, has long been known to possess significant biopharmaceutical activities such as antioxidant and anticancer potential [65–67]. There has been an increased attraction in its application in recent times. However, its clinical application is limited due to its poor solubility in water which leads to very poor absorption in the GI tract and hence low bioavailability. Wei et al. [67] proposed a sterically stabilized nanoscale dispersion loaded with curcumin based on a nonionic colloidal system produced by employing ultrasound and solvent diffusion-evaporation. In order to generate this nanodispersion, authors have utilized an ultrasonic bath operating at 38 kHz in two stages for 10 min to induce homogenization and droplet size reduction by employing the nonionic surfactants Span 20 and polyoxyethylene (10) cetyl ether. A narrow particle size distribution was effectively achieved, and the largest possible negative zeta potential was obtained by carefully selecting the ratio of surfactants.

Development of pharmaceutical formulations is not only a time-consuming process but also involves tedious processing. Thus, process optimization considering a wide range of influencing factors is always considered carefully. In the above study, the authors optimized the process conditions using Response Surface Optimization (RSM) and Box–Behnken Design (BBD) [67]. By this, they were able to obtain a nanodispersion with spherical droplets with a narrow distribution that could achieve a controlled release profile over 72 h.

Curcumin-loaded micelles (spherical entities formed by the self-aggregation of surfactant molecules in colloidal solution) have also been generated using ultrasonication [67]. The micelles produced had an average diameter of 20 nm with a polydispersity index (PDI) of 0.267. They were obtained using a surfactant with a HLB of 9.46 (mixture of Span 20 and Brij 56 surfactants), a curcumin-to-water weight ratio of 0.7:1 and surfactant-to-water weight ratio of 0.11:1. An increase in the PDI, which led to destabilisation of the nanoemulsion droplets, could have been due to droplet aggregation. This necessitates the usage of the right combination of surfactants and/or co-surfactants that can form a thick steric barrier against droplet coalescence.

3.1.2 Controlled Delivery of Bioactives Using Double Emulsions

RSM and central composite design (CCD) have been utilized to understand the impact of emulsion composition variables and ultrasonic operating parameters on the properties of aspirin-loaded nanoemulsions. Aspirin, a nonsteroidal anti-inflammatory drug (NSAID), is recommended as a nonprescription drug to relieve pain and inflammation. However, its long-term usage leads to various complications such as GI intolerance, stomach irritation and bleeding. These issues provide demand for newer formulations that can reduce the dosage required for effective treatment, to help minimize the side effects. Aspirin-incorporated O/W nanoemulsions were ultrasonically generated using a nonionic polyethoxylated surfactant (Cremophore EL) together with a co-surfactant (Transcutol HP) that enhances the solubility and hence bioavailability of the oil phase [68]. This study again proved the ability of ultrasound to produce small droplet with an average diameter in the range of 200–300 nm with a PDI of 0.3. Besides O/W nanoemulsions, water-in-oil-in-water (W/O/W) double emulsions [69] have been generated using ultrasonication. The anti-inflammatory and analgesic activities of both O/W and W/O/W nanoemulsions were determined using model in vitro trials [68], e.g. reduction of inflammation using λ-carrageenan-induced paw oedema model, acetic acid-induced writhing response and hot plate assays. The outcome of the above studies clearly indicated a pronounced improvement in the anti-inflammatory and analgesic activities compared to a conventional dosage of aspirin in the form of a simple suspension in water.

In another study, the osmotic behaviour of ultrasonically prepared nanomultiple emulsions incorporating aspirin was examined by changing the concentrations of glucose both in the inner and outer aqueous phases [70]. The role of gelatin in preventing the inter-droplet coalescence by forming an interfacial rigid film was also examined. This study revealed that the presence of glucose in the inner aqueous phase increased the average droplet size owing to swelling as well as due to an increase in the inner viscosity. An increase to the inner viscosity means that it is more resistant to shear by applied ultrasound, hence resulting in formation of larger internal droplet size.

To understand the energy efficiency, ultrasound has been compared with an air-driven microfluidizer to induce nanoemulsions using aspirin as a model drug [71]. The effects of process variables including prehomogenization using a rotor–stator high-speed mixer (Ultra-Turrax) and the extent of drug loading on the mean droplet diameter and size distribution of droplets were examined. The results indicated that both techniques were comparable, generating droplets in the desired size range of 150–170 nm. In the case of sonication, the obtained droplet size was dependent on prehomogenization and aspirin loading, while these factors had only a marginal effect in the case of the microfluidizer. More importantly, it has been noted that to achieve a similar minimum (aspirin-loaded) droplet diameter of around 160 nm, ultrasound was 18 times more energy-efficient than the microfluidizer, although the latter gave better physicochemical stability. However, it should be noted that this value has to be looked at cautiously as it is system-specific.

Recently, Alzorqi et al. [72] optimized the formulation of palm oil (as a dispersed phase)-based O/W nanoemulsions for the incorporation of β-D-glucan polysaccharides using CCD. Polysaccharides of β-D-glucan configuration, which are well known for their antioxidant activities, were extracted from a locally available mushroom, Ganoderma lucidum. Compared to high-speed rotor–stator mixing using an Ultra-Turrax system, the ultrasonic emulsification process generated smaller droplet sizes with narrower size distributions and greater stability, in a shorter period of time. An enhancement in the antioxidant activity was also observed in the ultrasound-induced nanoemulsion formulations.

Overall, the above studies clearly demonstrate the potential of employing low-frequency ultrasound to generate nanoemulsions or nanodispersions incorporated with active pharmaceutical/phytochemical components. Owing to the advantages of these novel nanoformulations produced using ultrasound, it is expected that they will play a key role as drug delivery systems in the near future.

3.2 Applications in Food Processing

3.2.1 Emulsions for the Dairy Industry

Milk proteins, which consist of caseins and whey proteins, are large molecular surfactant species that help stabilize the butter fat present in homogenized milks [73]. Whey proteins, available as powders in concentrate and isolate forms, and sodium caseinate are often used emulsifiers in the food industry. The application of ultrasound to milk proteins has been shown to result in partial denaturation of proteins (less than 1%) [45] leading to increased hydrophobicity that improves surface activity, as well as disrupting protein aggregates to form smaller particles.

Shanmugam and Ashokkumar [45] reported on the use of ultrasonic emulsification to create stable food-based emulsions of flaxseed oil directly in skim milk, using the naturally present milk proteins. The advantage of ultrasonic emulsification was showcased in this study, as a rotor–stator mixer (Ultra-Turrax mixer) operated with an equivalent energy input was unable to create nanoemulsions with comparable stability (Fig. 3.1). The emulsion droplet sizes produced using ultrasonication were also notably smaller when created using a high-speed rotor–stator (Ultra-Turrax) mixing at an equivalent energy density (Table 3.1). This result highlighted the potential importance of cavitation (which is present during ultrasonication but absent during rotor–stator mixing) to the emulsion stabilization process in protein systems. In this case, the intense energy (high temperature and pressure) evolved from cavitation hot spots can partially denature the skim milk proteins in the system, improving their ability to interface at the formed oil droplet interfaces (Table 3.2).

Leong et al. used this principle to reduce the overall surfactant usage in the formation of W/O/W-type double emulsions [46]. Native milk proteins were used exclusively to stabilize the external droplet interface, while a combined protein/surfactant system was used to stabilize the interface of the internalized droplets (Fig. 3.2). These double emulsions were able to stably encapsulate an inner aqueous phase with an encapsulation efficiency of between 30 and 100%.

Ultrasonication can also be effectively used as a simultaneous homogenization/emulsification and pasteurization technique for dairy processing. Thermosonication of milks, which is the application of ultrasound in combination with heating, has been used to effectively reduce whole milk droplets with a D[3,2] size of 2 μm to 0.5 μm (at 70 °C for 70 s). Bacterial reduction can also be achieved, with a 5 log reduction reported after 10 min of thermosonication treatment [80]. A 0.7 log reduction was achieved after the same duration when using heating alone. Due to the additional effect of acoustic cavitation, the heating requirement is reduced, meaning that pasteurization can be achieved with reduced temperatures or reduced time. This is of particular interest as a nonthermal (low-heat) preservation technique for dairy products. Reduced heat treatments for milk pasteurization have potential to be used to create milks with improved flavour and nutritional quality.

3.2.2 Antimicrobial Efficacy of Essential Oil Nanoemulsions

Essential oils have a wide range of beneficial properties including high nutritional value [81], flavour and antimicrobial properties [4], making them a prime candidate for incorporation in functional foods. Essential oils can be difficult to incorporate stably into aqueous-based food products due to the hydrophobic nature of the oils, but can be stably incorporated into food systems as nanoemulsions.

Ultrasonication has been used as an effective method to create stable nanoemulsions using basil [75], orange peel [3, 82], annatto seed [81] and *Thymus daenensis* [4] essential oils. In addition to being shelf-stable for many months without phase separation, these nanoemulsions were generally found to possess good-to-excellent antimicrobial activity. Sugumar et al. [79] used ultrasonically prepared orange oil nanoemulsions to control yeast viability in apple juice. After dilution in apple juice, the nanoemulsion rendered 100% of the yeast cells present in the juice to become unviable after 48 h. SEM images of yeast cells prior to and after inactivation by the nanoemulsions are depicted in Fig. 3.3.

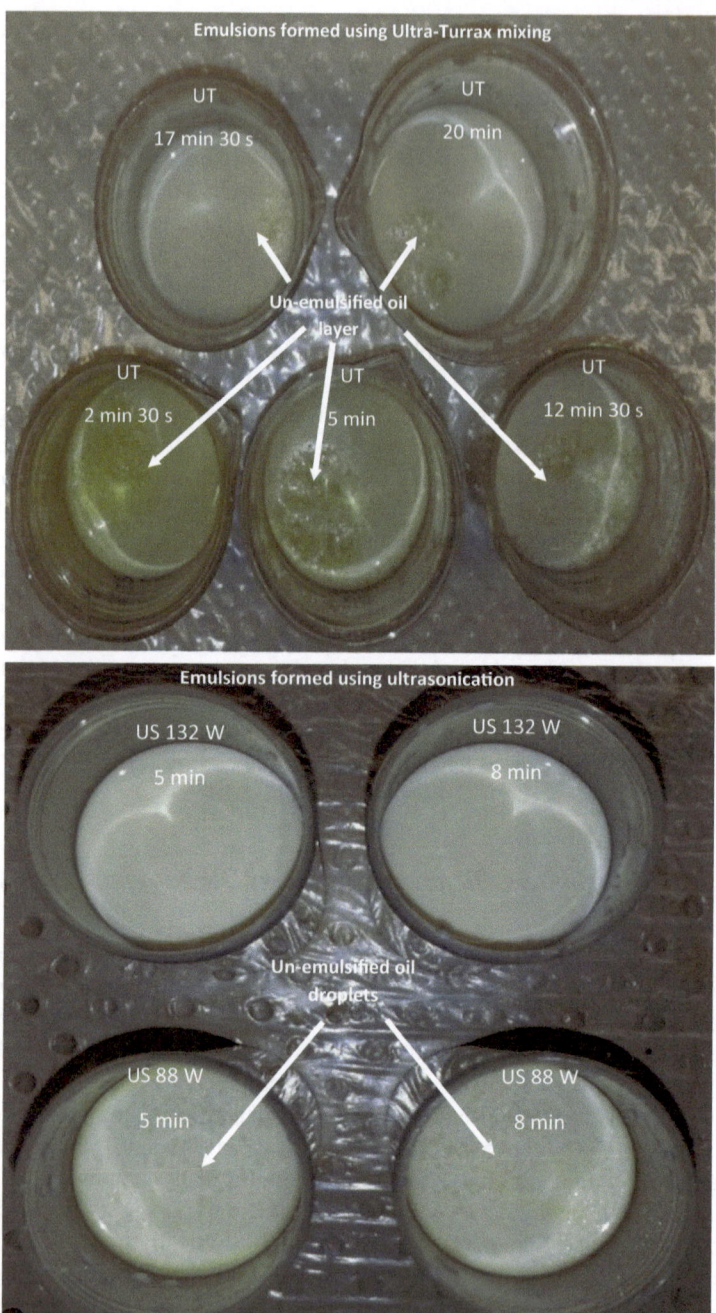

Fig. 3.1 Comparison of Ultra-Turrax (UT) and ultrasonication (US) of flaxseed oil into skim milk. Adapted from Shanmugam and Ashokkumar [45], Copyright 2014, with permission from Elsevier

Table 3.1 Selection of food and pharmaceutical-based applications in which ultrasonic emulsification has proven efficacy

Oil phase	Emulsifier system	Function	Comments	References
Olive oil	Whey protein concentrate, xanthan gum, guar gum and locust bean gum	Delivery vehicle	Hydrocolloids used to control viscosity/stability of ultrasonically formed emulsions	Kaltsa et al. [74]
Lemon grass essential oil	Tween 80	Flavour; Antimicrobial	Translucent nanoemulsions formed	Salvia-Trujillo et al. [2]
Basil oil	Tween 80	Antioxidant; Antimicrobial	Bactericidal activity against *Escherichia coli*	Ghosh et al. [75]
Fish oil	Tween 80, Span 80	Antioxidant; Delivery vehicle	Faster intestinal absorption of lipids	Kumar Dey et al. [76]
D-limonene	Span 80, Brij 98/ethylene glycol	Flavour; Antioxidant; Delivery vehicle	Translucent nanoemulsions formed	Li and Chiang [39]
Lemon oil	Tween 80, PG8	Flavour; Delivery vehicle	Comparison with thermal treatments	Rao and McClements [77]
Sunflower oil, canola oil	Tween 80, Span 80, SDS	Delivery vehicle	Eye-clear nanoemulsions formed	Leong et al. [8]
Flaxseed oil	Skim milk	Delivery vehicle	Ultrasonication produced more stable emulsions compared with Ultra-Turrax at matched energy delivery	Shanmugam and Ashokkumar [45]
Flaxseed oil	Egg lecithin, ethanol	Curcumin delivery (anticancer)	Delivery vehicle for anticancer in ovarian cells	Ganta and Amiji [65]
Sunflower oil	Skim milk, Span 80	Delivery vehicle; Fat reduction	Double emulsions produced using ultrasonication in skim milk	Leong et al. [46]
Linseed oil	Tween 40	Drug delivery	Contactless ultrasonic system developed	Freitas et al. [60]
Soybean oil	Tween 80	Delivery vehicle	Scale-up formation of translucent nanoemulsions	Peshkovsky et al. [31]

(continued)

Table 3.1 (continued)

Oil phase	Emulsifier system	Function	Comments	References
Mustard oil	Span 80, Tween 80	Delivery vehicle	Process parameters investigated	Carpenter and Saharan [78]
Orange oil	Tween 80	For food preservation	Mitigation of yeast viability in apple juice	Sugumar et al. [79]
Nigella sativa L. essential oil	Tween 80	Anticancer; Antioxidant; Antimicrobial	Nanoemulsions demonstrated anticancer efficacy on breast cancer cells	Periasamy et al. [61]

Table 3.2 Comparison of volume-weighted mean diameter (D(4,3)) of ultrasonication (US) and Ultra-Turrax (UT) processing with matched energy density of flaxseed oil in skim milk

US processing time (min:s)	Equivalent UT processing time (min:s)	D(4,3) of US (μm)	D(4,3) of UT (μm)
1:00	2:30	1.38	3.30
2:00	5:00	0.83	2.27
5:00	12:30	0.48	1.54
7:00	17:30	0.39	2.20
8:00	20:00	0.40	1.48

Adapted with permission from Shanmugam and Ashokkumar [45], Copyright 2014, with permission from Elsevier

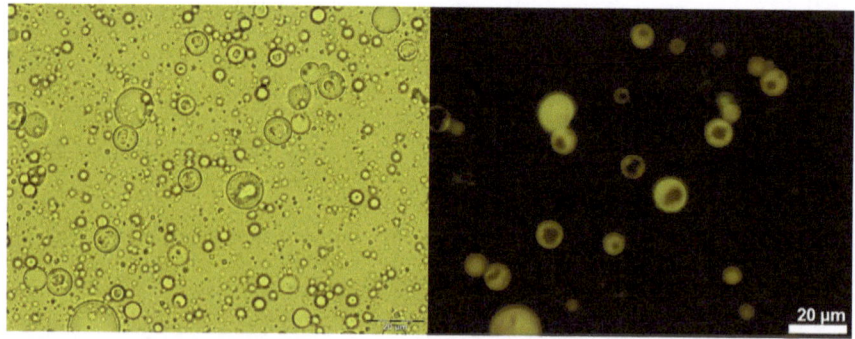

Fig. 3.2 Double emulsion of skim milk encapsulated within sunflower oil dispersed within skim milk, formed using ultrasonication. Reprinted from Leong et al. [46], Copyright 2017, with permission from Elsevier

The antimicrobial activity for nanoemulsions of these essential oils was found to be superior to the un-emulsified oil in a pure form [4]. It was speculated that the increased activity was due to the nanoemulsions allowing the essential oils to come nearer to the bacterial cell membrane interface, enabling more effective disruption of the phospholipid bilayer. The activity of bioactive components may be negatively

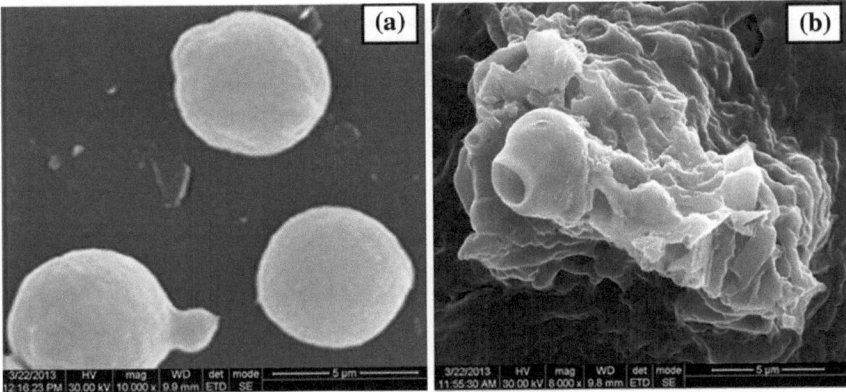

Fig. 3.3 Scanning electron microscope image of *Saccharomyces cerevisiae* cells **a** without nanoemulsion exposure, **b** with orange oil nanoemulsion exposure. Sourced from the open access reference by Sugumar et al. [79]

affected by the formation of radicals during ultrasonication [83]. Fortunately, treatment times required for emulsion formation are typically short, which limits the amount of degradation that occurs.

3.3 Future Trends and Outlook

The growing trend for developing functional foods will be supported by improvements in the efficiency of producing nanoemulsions. Ultrasonics as an emulsification technique is highly comparable to conventional methods such as high-pressure homogenization in regard to energy efficiency and effectiveness. Beyond this, ultrasonication can provide novel benefits such as modification of functional properties, caused by the forces generated by acoustic cavitation, in addition to simple shearing efficacy. These novel benefits will assist in the development of new food and pharmaceutical products.

The development of multiple emulsions for encapsulation of bioactives or fat displacement in food and pharmaceutical products shows excellent promise for various applications. However, they have yet to be commercially implemented into widely available products due to the high surfactant requirements and instability problems. The capability of ultrasound to produce multiple emulsions with tailored droplet sizes and morphology, together with its ability to improve the efficacy of natural emulsifiers, will enable fast-tracked development of these emulsions into viable products for human consumption.

Ongoing research is currently leading towards improved understanding of how to achieve more efficient and effective production of emulsions in general. As cavitation is a common mechanism across high-shear techniques, the ability to formulate emulsions and design processing conditions based on acoustic cavitation principles will complement and add to the existing knowledge of conventionally available emulsification equipment in industry. Continual development will reduce the capital cost of ultrasonic equipment, making investment by food and pharmaceutical companies in the technology a more attractive proposition.

References

1. I. Kralova, J. Sjöblom, Surfactants used in food industry: a review. J. Dispersion Sci. Technol. **30**, 1363–1383 (2009)
2. L. Salvia-Trujillo, A. Rojas-Graü, R. Soliva-Fortuny et al., Physicochemical characterization of lemongrass essential oil–alginate nanoemulsions: effect of ultrasound processing parameters. Food Bioprocess Tech. **6**, 2439–2446 (2013)
3. A.M. Hashtjin, S. Abbasi, Optimization of ultrasonic emulsification conditions for the production of orange peel essential oil nanoemulsions. J. Food Sci. Tech **52**, 2679–2689 (2015)
4. R. Moghimi, L. Ghaderi, H. Rafati et al., Superior antibacterial activity of nanoemulsion of *Thymus daenensis* essential oil against *E. coli*. Food Chem **194**, 410–415 (2016)
5. H. Lamba, K. Sathish, L. Sabikhi, Double emulsions: emerging delivery system for plant bioactives. Food Bioprocess Tech **8**, 709–728 (2015)
6. S.M. Jafari, Y. He, B. Bhandari, Production of sub-micron emulsions by ultrasound and microfluidization techniques. J. Food Eng **82**, 478–488 (2007)
7. S. Kentish, T. Wooster, M. Ashokkumar et al., The use of ultrasonics for nanoemulsion preparation. Innov. Food Sci. Emerg **9**, 170–175 (2008)
8. T.S.H. Leong, T.J. Wooster, S.E. Kentish et al., Minimising oil droplet size using ultrasonic emulsification. Ultrason. Sonochem **16**, 721–727 (2009)
9. C. Solans, P. Izquierdo, J. Nolla et al., Nano-emulsions. Curr. Opin. Colloid. **10**, 102–110 (2005)
10. D. Kilcast, S. Clegg, Sensory perception of creaminess and its relationship with food structure. Food Qual. Prefer **13**, 609–623 (2002)
11. D.J. Mela, K.R. Langley, A. Martin, Sensory assessment of fat content: effect of emulsion and subject characteristics. Appetite **22**, 67–81 (1994)
12. S.M. Jafari, Y. He, B. Bhandari, Nano-emulsion production by sonication and microfluidization—a comparison. Int. J. Food Prop **9**, 475–485 (2006)
13. B.E. Noltingk, E.A. Neppiras, Cavitation produced by ultrasonics. Proc. Phys. Soc. B **63**, 674 (1950)
14. T. Leong, M. Ashokkumar, S. Kentish, The fundamentals of power ultrasound—a review. Acoust. Aust **39**, 54–63 (2011)
15. M.P. Brenner, S. Hilgenfeldt, D. Lohse, Single-bubble sonoluminescence. Rev. Mod. Phys. **74**, 425 (2002)
16. D.J. Flannigan, K.S. Suslick, Plasma formation and temperature measurement during single-bubble cavitation. Nature **434**, 52–55 (2005)
17. J. Lighthill, Acoustic streaming. J. Sound Vib. **61**, 391–418 (1978)
18. C.F. Naudé, A.T. Ellis, On the mechanism of cavitation damage by nonhemispherical cavities collapsing in contact with a solid boundary. J. Fluids Eng. **83**, 648–656 (1961)
19. K.S. Suslick, Sonochemistry. Science **247**, 1439–1445 (1990)

© The Author(s) under exclusive licence to Springer International Publishing AG, part of Springer Nature 2018
T. S. H. Leong et al., *Ultrasonic Production of Nano-emulsions for Bioactive Delivery in Drug and Food Applications*, Ultrasound and Sonochemistry, https://doi.org/10.1007/978-3-319-73491-0

20. S.D. Collyer, F. Davis, S.P. Higson, Sonochemically fabricated microelectrode arrays for use as sensing platforms. Sens.-Basel 10, 5090–5132 (2010)
21. R.W. Wood, A.L. Loomis, XXXVIII. The physical and biological effects of high-frequency sound-waves of great intensity. Philos. Mag. Ser. 7(4), 417–436 (1927)
22. E.B. Flint, K.S. Suslick, The temperature of cavitation. Science 253, 1397–1399 (1991)
23. O. Behrend, K. Ax, H. Schubert, Influence of continuous phase viscosity on emulsification by ultrasound. Ultrason. Sonochem. 7, 77–85 (2000)
24. A. Håkansson, L. Fuchs, F. Innings et al., Velocity measurements of turbulent two-phase flow in a high-pressure homogenizer model. Chem. Eng. Commun. 200, 93–114 (2013)
25. S. Schultz, G. Wagner, K. Urban et al., High-pressure homogenization as a process for emulsion formation. Chem. Eng. Technol. 27, 361–368 (2004)
26. P. Walstra, Principles of emulsion formation. Chem. Eng. Sci. 48, 333–349 (1993)
27. Y.Y.J. Zuo, P. Hébraud, Y. Hemar et al., Quantification of high-power ultrasound induced damage on potato starch granules using light microscopy. Ultrason. Sonochem. 19, 421–426 (2012)
28. S. Abbas, K. Hayat, E. Karangwa et al., An overview of ultrasound-assisted food-grade nanoemulsions. Food Eng. Rev. 5, 139–157 (2013)
29. J. Canselier, H. Delmas, A. Wilhelm et al., Ultrasound emulsification—an overview. J. Dispersion Sci. Technol. 23, 333–349 (2002)
30. M. Ashokkumar, R. Bhaskaracharya, S. Kentish et al., The ultrasonic processing of dairy products—an overview. Dairy Sci. Technol. 90, 147–168 (2010)
31. A.S. Peshkovsky, S.L. Peshkovsky, S. Bystryak, Scalable high-power ultrasonic technology for the production of translucent nanoemulsions, Chem. Eng. Process. Process Intensification, 69, 77–82 (2013)
32. M. Ashokkumar, *Theoretical and Experimental Sonochemistry Involving Inorganic Systems* (Springer Science & Business Media, Germany, 2010)
33. A. Gupta, V. Narsimhan, T.A. Hatton et al., Kinetics of the change in droplet size during nanoemulsion formation. Langmuir 32, 11551–11559 (2016)
34. A. Gupta, H.B. Eral, T.A. Hatton et al., Controlling and predicting droplet size of nanoemulsions: scaling relations with experimental validation. Soft Matter 12, 1452–1458 (2016)
35. J. Hinze, Fundamentals of the hydrodynamic mechanism of splitting in dispersion processes. AIChE J. 1, 289–295 (1955)
36. V.M. Tikhomirov, On the breakage of drops in a turbulent flow. in Selected Works of AN Kolmogorov. Springer, Dordrecht. 339–343 (1991)
37. S. Tcholakova, N. Vankova, N.D. Denkov et al., Emulsification in turbulent flow: 3. Daughter drop-size distribution. J. Colloid Interface Sci. 310, 570–589 (2007)
38. N. Vankova, S. Tcholakova, N.D. Denkov et al., Emulsification in turbulent flow: 2. Breakage rate constants. J. Colloid Interface Sci. 313, 612–629 (2007)
39. P.-H. Li, B.-H. Chiang. Process optimization and stability of D-limonene-in-water nanoemulsions prepared by ultrasonic emulsification using response surface methodology. Ultrason. Sonochem. 19, 192–197 (2012)
40. O. Behrend, H. Schubert, Influence of hydrostatic pressure and gas content on continuous ultrasound emulsification. Ultrason. Sonochem. 8, 271–276 (2001)
41. T.G. Leighton, *The Acoustic Bubble* (Academic Press, San Diego, 1994)
42. T. Kimura, T. Sakamoto, J.-M. Leveque et al., Standardization of ultrasonic power for sonochemical reaction. Ultrason. Sonochem. 3, S157–S161 (1996)
43. C. Bondy, K. Söllner, On the mechanism of emulsification by ultrasonic waves. T. Faraday Soc. 31, 835–843 (1935)
44. P. Marie, J. Perrier-Cornet, P. Gervais, Influence of major parameters in emulsification mechanisms using a high-pressure jet. J. Food Eng. 53, 43–51 (2002)
45. A. Shanmugam, M. Ashokkumar, Ultrasonic preparation of stable flax seed oil emulsions in dairy systems—physicochemical characterization. Food Hydrocolloid 39, 151–162 (2014)

46. T.S. Leong, M. Zhou, N. Kukan et al., Preparation of water-in-oil-in-water emulsions by low frequency ultrasound using skim milk and sunflower oil. Food Hydrocolloid **63**, 685–695 (2017)

47. C. Perks, M. Piatko, T. Aurand, in *Whippable Food Product with Improved Stability*, Google Patents (2008)

48. D.J. Flannigan, K.S. Suslick, Molecular and atomic emission during single-bubble cavitation in concentrated sulfuric acid. Acoust. Res. Lett. Online **6**, 157–161 (2005)

49. K. Nakabayashi, F. Amemiya, T. Fuchigami et al., Highly clear and transparent nanoemulsion preparation under surfactant-free conditions using tandem acoustic emulsification. Chem. Commun. **47**, 5765–5767 (2011)

50. K. Kamogawa, G. Okudaira, M. Matsumoto et al., Preparation of oleic acid/water emulsions in surfactant-free condition by sequential processing using midsonic-megasonic waves. Langmuir **20**, 2043–2047 (2004)

51. M. Ashokkumar, D. Sunartio, S. Kentish et al., Modification of food ingredients by ultrasound to improve functionality: a preliminary study on a model system. Innov. Food Sci. Emerg. **9**, 155–160 (2008)

52. S. Koda, T. Kimura, T. Kondo et al., A standard method to calibrate sonochemical efficiency of an individual reaction system. Ultrason. Sonochem. **10**, 149–156 (2003)

53. T.J. Mason, A.J. Cobley, J.E. Graves et al., New evidence for the inverse dependence of mechanical and chemical effects on the frequency of ultrasound. Ultrason. Sonochem. **18**, 226–230 (2011)

54. P. Juliano, A.E. Torkamani, T. Leong et al., Lipid oxidation volatiles absent in milk after selected ultrasound processing. Ultrason. Sonochem. **21**, 2165–2175 (2014)

55. J. Chandrapala, T. Leong, Ultrasonic processing for dairy applications: recent advances. Food Eng. Rev. **7**(2), 143–158 (2015)

56. F. Chemat, I. Grondin, P. Costes et al., High power ultrasound effects on lipid oxidation of refined sunflower oil. Ultrason. Sonochem. **11**, 281–285 (2004)

57. J. Chandrapala, B. Zisu, S. Kentish et al., Influence of ultrasound on chemically induced gelation of micellar casein systems. J. Dairy Res. **80**, 138–143 (2013)

58. J. Chandrapala, G. Martin, B. Zisu et al., The effect of ultrasound on casein micelle integrity. J. Dairy Sci. **95**, 6882–6890 (2012)

59. R. Mawson, M. Rout, G. Ripoll et al., Production of particulates from transducer erosion: implications on food safety. Ultrason. Sonochem. **21**(6), 2122–2130 (2014)

60. S. Freitas, G. Hielscher, H.P. Merkle et al., Continuous contact-and contamination-free ultrasonic emulsification—a useful tool for pharmaceutical development and production. Ultrason. Sonochem. **13**, 76–85 (2006)

61. V.S. Periasamy, J. Athinarayanan, A.A. Alshatwi, Anticancer activity of an ultrasonic nanoemulsion formulation of *Nigella sativa* L. essential oil on human breast cancer cells. Ultrason. Sonochem. **31**, 449–455 (2016)

62. T.S. Leong, G.J. Martin, M. Ashokkumar, Ultrasonic encapsulation—a review. Ultrason. Sonochem. **35**, 605–614 (2016)

63. M.N. Martinez, G.L. Amidon, A mechanistic approach to understanding the factors affecting drug absorption: a review of fundamentals. J. Clin. Pharmacol. **42**, 620–643 (2002)

64. M. Sivakumar, S.Y. Tang, K.W. Tan, Cavitation technology—a greener processing technique for the generation of pharmaceutical nanoemulsions. Ultrason. Sonochem. **21**, 2069–2083 (2014)

65. S. Ganta, M. Amiji, Coadministration of paclitaxel and curcumin in nanoemulsion formulations to overcome multidrug resistance in tumor cells. Mol. Pharm. **6**, 928–939 (2009)

66. K.W. Tan, S.Y. Tang, R. Thomas et al., Curcumin-loaded sterically stabilized nanodispersion based on non-ionic colloidal system induced by ultrasound and solvent diffusion-evaporation. Pure Appl. Chem. **88**, 43–60 (2016)

67. T.K. Wei, S. Manickam, Response surface methodology, an effective strategy in the optimization of the generation of curcumin-loaded micelles. Asia-Pacific J. Chem. Eng. **7** (S1) (2012)

68. S.Y. Tang, M. Sivakumar, A.M.-H. Ng et al., Anti-inflammatory and analgesic activity of novel oral aspirin-loaded nanoemulsion and nano multiple emulsion formulations generated using ultrasound cavitation. Int. J. Pharm. **430**, 299–306 (2012)

69. S.Y. Tang, M. Sivakumar, Design and evaluation of aspirin-loaded water-in-oil-in-water submicron multiple emulsions generated using two-stage ultrasonic cavitational emulsification technique. Asia-Pacific J. Chem. Eng. **7**, S145–S156 (2012)

70. S.Y. Tang, M. Sivakumar, B. Nashiru, Impact of osmotic pressure and gelling in the generation of highly stable single core water-in-oil-in-water (W/O/W) nano multiple emulsions of aspirin assisted by two-stage ultrasonic cavitational emulsification. Colloid Surf. B **102**, 653–658 (2013)

71. S.Y. Tang, P. Shridharan, M. Sivakumar, Impact of process parameters in the generation of novel aspirin nanoemulsions–comparative studies between ultrasound cavitation and microfluidizer. Ultrason. Sonochem. **20**, 485–497 (2013)

72. I. Alzorqi, M.R. Ketabchi, S. Sudheer et al., Optimization of ultrasound induced emulsification on the formulation of palm-olein based nanoemulsions for the incorporation of antioxidant β-D-glucan polysaccharides. Ultrason. Sonochem. **31**, 71–84 (2016)

73. M.-C. Michalski, F. Michel, D. Sainmont et al., Apparent ζ-potential as a tool to assess mechanical damages to the milk fat globule membrane. Colloid Surf. B **23**, 23–30 (2002)

74. O. Kaltsa, C. Michon, S. Yanniotis et al., Ultrasonic energy input influence on the production of sub-micron o/w emulsions containing whey protein and common stabilizers. Ultrason. Sonochem. **20**, 881–891 (2013)

75. V. Ghosh, A. Mukherjee, N. Chandrasekaran, Ultrasonic emulsification of food-grade nanoemulsion formulation and evaluation of its bactericidal activity. Ultrason. Sonochem. **20**, 338–344 (2013)

76. T. Kumar Dey, S. Ghosh, M. Ghosh et al., Comparative study of gastrointestinal absorption of EPA & DHA rich fish oil from nano and conventional emulsion formulation in rats. Food Res. Int. **49**, 72–79 (2012)

77. J. Rao, D.J. McClements, Formation of flavor oil microemulsions, nanoemulsions and emulsions: influence of composition and preparation method. J. Agr. Food Chem. **59**, 5026–5035 (2011)

78. J. Carpenter, V.K. Saharan, Ultrasonic assisted formation and stability of mustard oil in water nanoemulsion: effect of process parameters and their optimization. Ultrason. Sonochem. **35**, 422–430 (2017)

79. S. Sugumar, S. Singh, A. Mukherjee et al., Nanoemulsion of orange oil with non ionic surfactant produced emulsion using ultrasonication technique: evaluating against food spoilage yeast. Appl. Nanosci. **6**, 113–120 (2016)

80. D. Bermúdez-Aguirre, R. Mawson, G. Barbosa-Cánovas, Microstructure of fat globules in whole milk after thermosonication treatment. J. Food Sci. **73**, 325–332 (2008)

81. E.K. Silva, M.T.M. Gomes, M.D. Hubinger et al., Ultrasound-assisted formation of annatto seed oil emulsions stabilized by biopolymers. Food Hydrocolloid **47**, 1–13 (2015)

82. A.M. Hashtjin, S. Abbasi, Nano-emulsification of orange peel essential oil using sonication and native gums. Food Hydrocolloid **44**, 40–48 (2015)

83. A.C. Soria, M. Villamiel, Effect of ultrasound on the technological properties and bioactivity of food: a review. Trends Food Sci. Tech. **21**, 323–331 (2010)